会飞的史前巨无霸

Prehistoric Flying Giants

[英] 露丝・欧文/著
刘颖/译

汉英对照
恐龙科普

江苏凤凰美术出版社

全家阅读小贴士

★ 每天空出大约10分钟来阅读。

★ 找个安静的地方坐下，集中注意力。关掉电视、音乐和手机。

★ 鼓励孩子们自己拿书和翻页。

★ 开始阅读前，先一起看看书里的图画，说说你们看到了什么。

★ 如果遇到不认识的单词，先问问孩子们首字母如何发音，再带着他们读完整句话。

★ 很多时候，通过首字母发音并听完整句话，孩子们就能猜出单词的意思。书里的图画也能起到提示的作用。

最重要的是，感受一起阅读的乐趣吧！

扫码听本书英文

Tips for Reading Together

- Set aside about 10 minutes each day for reading.
- Find a quiet place to sit with no distractions. Turn off the TV, music and screens.
- Encourage the child to hold the book and turn the pages.
- Before reading begins, look at the pictures together and talk about what you see.
- If the child gets stuck on a word, ask them what sound the first letter makes. Then, you read to the end of the sentence.
- Often by knowing the first sound and hearing the rest of the sentence, the child will be able to figure out the unknown word. Looking at the pictures can help, too.

Above all enjoy the time together and make reading fun!

Contents 目录

一群恐龙正在吃植物。

突然，一头巨大的动物从它们头顶飞过。

是哈特兹哥翼龙，它在寻找猎物！

这群恐龙危险了。

Some dinosaurs were feeding on plants.

Then a huge animal flew over them.

It was Hatzegopteryx and it was looking for a meal!

The dinosaurs were in danger.

哈特兹哥翼龙生活在大约7000万年前。
Hatzegopteryx lived about 70 million years ago.

认识哈特兹哥翼龙
Meet Hatzegopteryx

哈特兹哥翼龙是有史以来最大的飞行动物之一。它的翼展将近12米。

Hatzegopteryx was one of the biggest flying animals that ever lived. Its **wingspan** was almost 12 metres.

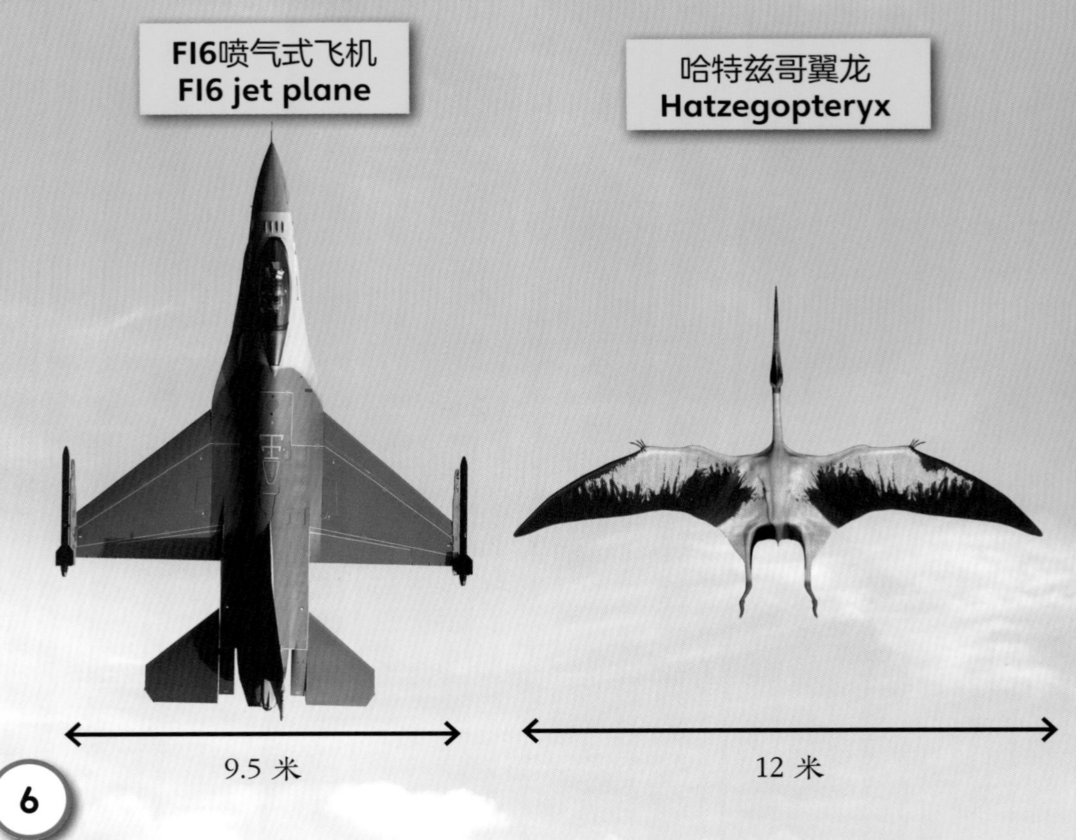

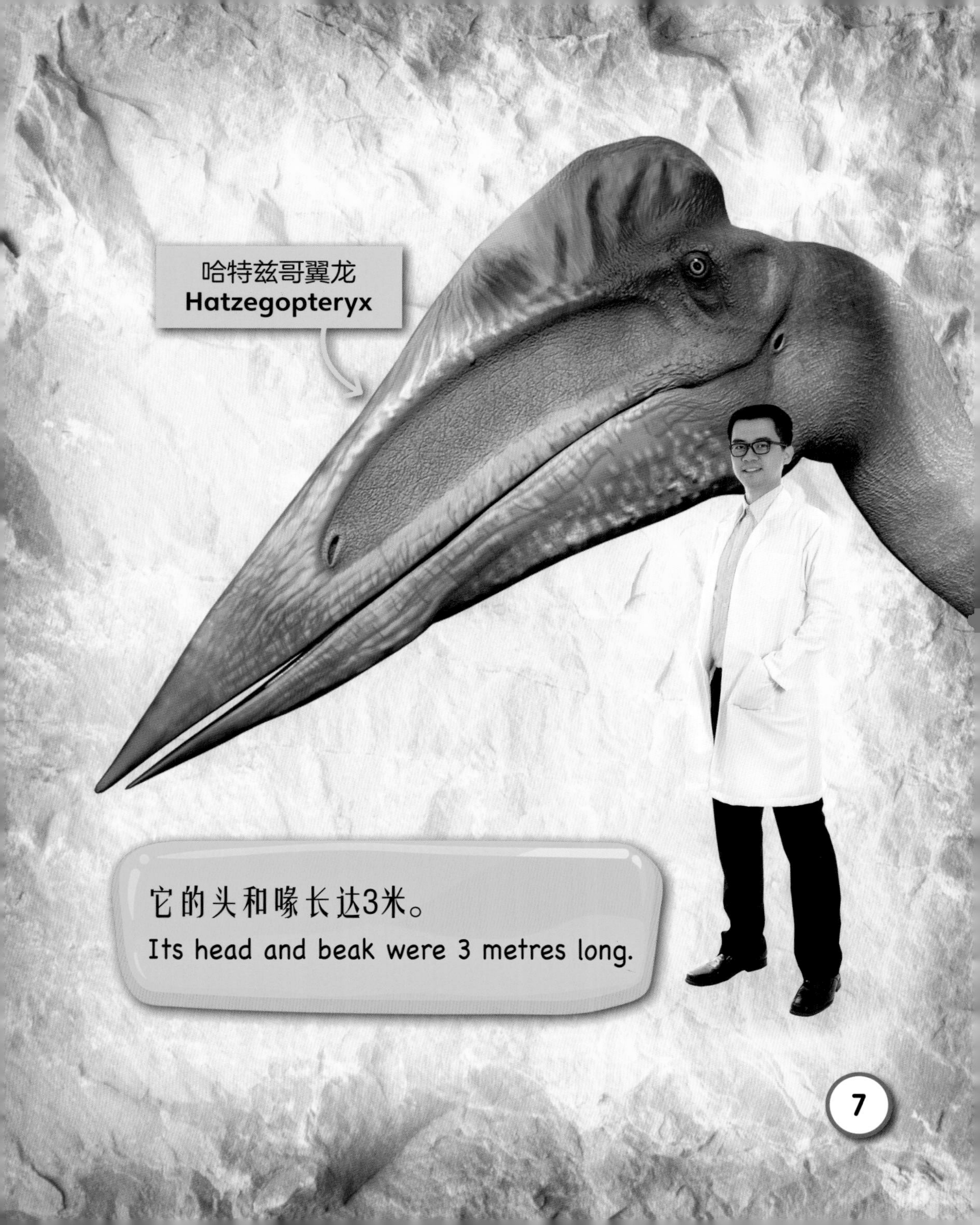

它的头和喙长达3米。

Its head and beak were 3 metres long.

寻找巨无霸
Finding a Giant

1991年，科学家发现了一些巨大的骨骼化石。他们认为这是恐龙的骨头。

In 1991 **scientists** found some big **fossil** bones.
They thought the bones were from a dinosaur.

红色的骨头是那些被发现的化石。
The red bones are the fossils.

哈特兹哥翼龙的骨骼
Hatzegopteryx skeleton

但他们新发现了一种巨大的飞行动物。

But they had found a new huge flying animal.

哈特兹哥翼龙比公交车还高。
Hatzegopteryx was taller than a bus.

哈特兹哥翼龙不是鸟，也不是恐龙。
它是一种翼龙。
翼龙是会飞的爬行动物。

Hatzegopteryx was not a bird or a dinosaur.
It was a pterosaur (TER-uh-sawr).
Pterosaurs were flying **reptiles**.

翼龙
pterosaur

现在有很多爬行动物，比如鳄鱼，但它们都不会飞。

There are reptiles, like crocodiles, around today but there are no flying reptiles.

翼龙的翅膀
Pterosaur Wings

翼龙的前肢末端各长着一只爪子。

A pterosaur had a hand at the end of each front leg.

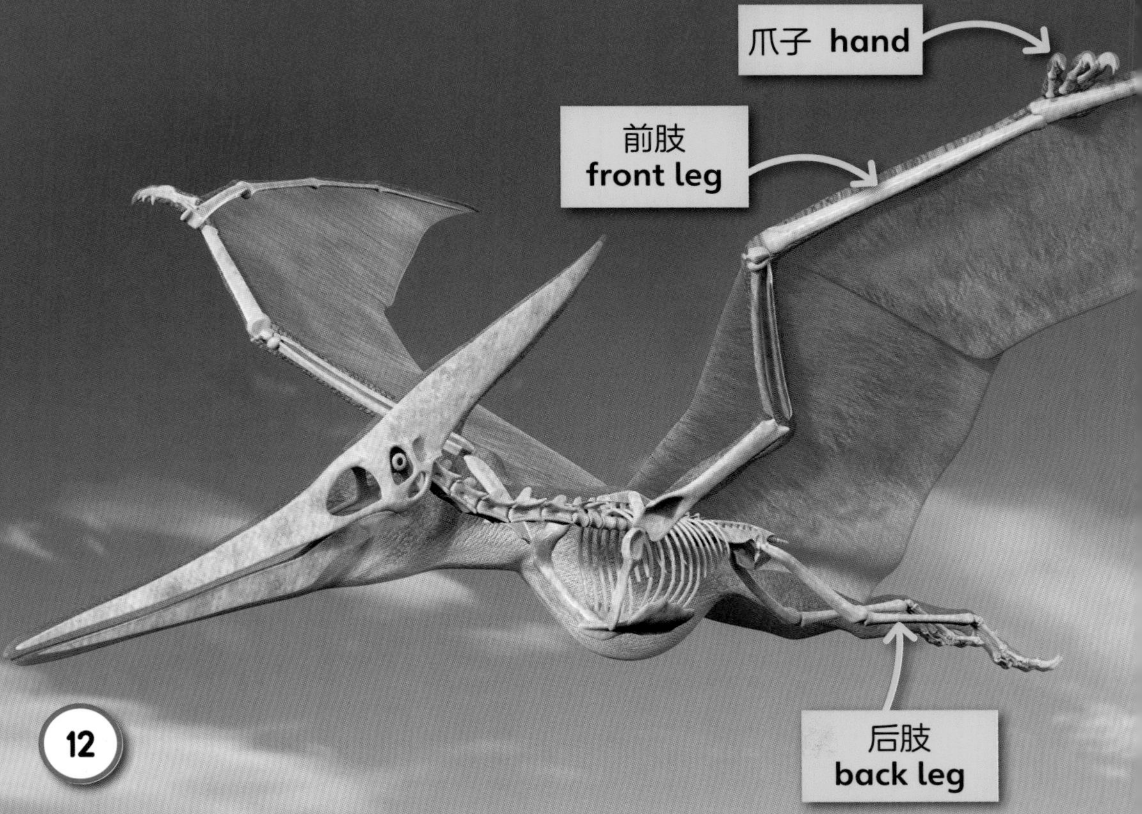

爪子 hand

前肢
front leg

后肢
back leg

每只爪子都有一根超长的指骨。

Each hand had one extra-long finger.

翅膀连接极长的指骨与后肢。

The wings joined the long fingers to the back legs.

飞翔和行走
Flying and Walking

翼龙的骨头是中空的，像管子一样。这使翼龙的体重更轻，有助于飞翔。

A pterosaur's bones were hollow, like tubes. This made the animal lighter to help it fly.

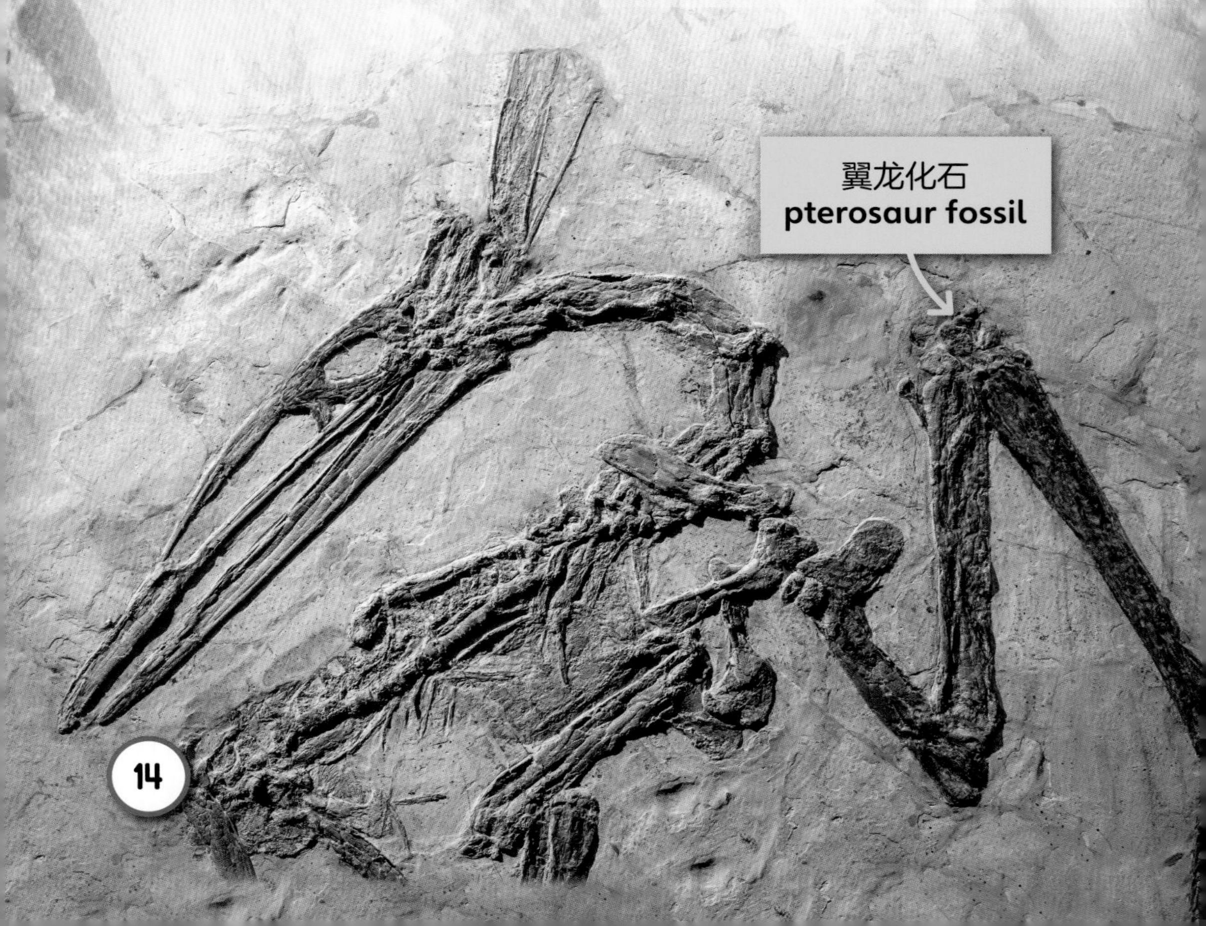

当翼龙降落时，它会收起翅膀，用手和脚行走。
When a pterosaur landed it would bend its wings and walk on its hands and feet.

庞大的猎手
Huge Hunters

哈特兹哥翼龙是食肉动物。

它能猎杀其他恐龙。

它用喙撕咬和刺死猎物。

Hatzegopteryx was a meat-eater.

It hunted dinosaurs.

It used its beak to bite and stab its **prey**.

哈特兹哥翼龙有一张巨大的嘴。
它能一口气吞下一只小恐龙。
Hatzegopteryx had a huge mouth.
It could swallow small dinosaurs in one go.

巨型翼龙
Enormous Pterosaurs

哈特兹哥翼龙并不是唯一的巨型翼龙。有一种名为阿氏翼龙的巨型翼龙和长颈鹿一样高！

Hatzegopteryx wasn't the only enormous pterosaur.
A pterosaur called Arambourgiania was as tall as a giraffe!

阿氏翼龙
Arambourgiania
(a-RAM-bore-gee-uh-NEE-a)

还有一种巨型翼龙是风神翼龙。
它长着非常锋利的喙。

Another huge pterosaur was Quetzalcoatlus.

It had a very sharp beak.

风神翼龙的同比例模型
a life-size model of Quetzalcoatlus (KWET-sa-COTE-lass)

四处飞行

Flying Around

风神翼龙没有羽毛。
它的身上长满了短毛。

Quetzalcoatlus did not have feathers.
Its body was covered with short hairs.

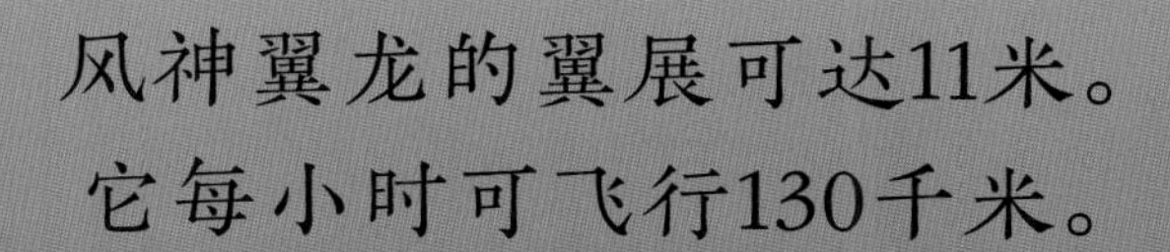

风神翼龙的翼展可达11米。
它每小时可飞行130千米。

Quetzalcoatlus had a wingspan of 11 metres.
It could fly at 130 kilometres per hour.

恐龙蛋化石
fossil eggs

同短吻鳄蛋一样，翼龙蛋的外壳也有弹性。
Pterosaur eggs had rubbery shells like alligator eggs.

短吻鳄蛋
alligator eggs

词汇表 Glossary

化石　fossil

存留在岩石中几百万年前的动物和植物的遗体。

The rocky remains of an animal or plant that lived millions of years ago.

猎物　prey

被其他动物猎食的动物。

Animals that are hunted and eaten by other animals.

爬行动物 reptiles

包括恐龙、翼龙和现代的蜥蜴、短吻鳄等动物。

Animals including dinosaurs, pterosaurs and modern animals such as lizards and alligators.

科学家 scientist

研究自然和世界的人。

A person who studies nature and the world.

翼展 wingspan

两侧翼尖之间的距离。

The distance between the tips of two wings.

恐龙小测验 Dinosaur Quiz

1. 哈特兹哥翼龙的头和喙有多大？
   How big was Hatzegopteryx's head and beak?

2. 哈特兹哥翼龙是鸟吗？
   Was Hatzegopteryx a bird?

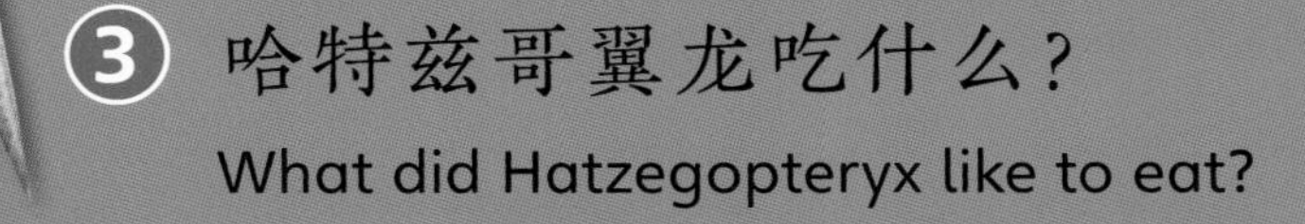

3. 哈特兹哥翼龙吃什么？
   What did Hatzegopteryx like to eat?

4. 阿氏翼龙有多高？
   How tall was Arambourgiania?

5. 假如你看见了正在飞行的风神翼龙，你会害怕吗？
   Would you be scared if you saw a Quetzalcoatlus in the sky?